INTERNATIONAL CENTRE FOR MECHANICAL SCIENCES

COURSES AND LECTURES - No. 179

HAYRETTIN KARDESTUNCER
UNIVERSITY OF STORRS, CONNECTICUT

MATRIX ANALYSIS
OF
DISCRETE ELASTIC SYSTEMS

COURSE HELD AT THE DEPARTMENT
OF MECHANICS OF SOLIDS
OCTOBER 1972

UDINE 1974

SPRINGER-VERLAG WIEN GMBH

This work is subject to copyright.

All rights are reserved,

whether the whole or part of the material is concerned

specifically those of translation, reprinting, re-use of illustrations,

broadcasting, reproduction by photocopying machine

or similar means, and storage in data banks.

© 1972 by Springer-Verlag Wien

Originally published by Springer-Verlag Wien New York in 1972

ISBN 978-3-211-81235-8   ISBN 978-3-7091-2910-4 (eBook)
DOI 10.1007/978-3-7091-2910-4

PREFACE

The analysis of discrete elastic systems (structures) is presented. Whether a system is discrete to begin with or is discretized in a certain fashion (idealization of continuum by finite elements), its analysis within the framework of matrix algebra has received ever-increasing popularity after the middle of this century. In order to prepare the audience to think in terms of entities, we have introduced a brief review of matrix algebra and coordinate transformations in Chapter 1. Thereafter, we go into the analysis of structures using stiffness and flexibility methods which are derived from Castigliano's first and second theorems respectively. Through such theorems, we clearly see the duality of these two well-known methods.

Since the purpose of analysis is to achieve an optimal design, the modification of systems after each analysis is very common. Therefore, in Chapter 4, we discuss how to modify the method of analysis in order to take advantage of the previous analysis. Such a feature, which did not exist in pre-matrix methods, allows a considerable amount of saving in machine time and labor.

Finally, once the analyst knows well how to play the game (analysis), he then looks into possibilities of playing the game more efficiently. Therefore, the analysis of large systems by the Method of Tearing and their replacement by equivalent systems are presented in Chapters 5 and 6 respectively.

It is indeed my great pleasure to express my gratitude to Professor Sobrero, the Secretary General of CISM, for providing me with such an excellent opportunity and to Professor Olszak, Rector of CISM, for making my stay in Udine pleasant and memorable. Furthermore, the kind and generous cooperation of the Editorial Board, headed by Dr. Buttazzoni, is greatly appreciated.

H. Kardestuncer

Udine, October 1972.

Chapter 1
MATRICES, SIMULTANEOUS EQUATIONS AND COORDINATE TRANSFORMATIONS

1.1. Matrices

Matrices are the set of m by n numbers arranged in a rectangular array with m rows and n columns. The order of matrices is determined by the number of rows and columns and normally written as mXn. If m = n, the matrix is called a square matrix of order n. If n = 1, it is called a column matrix or a vector.

There are three types of products in vectors:

$$\underset{\sim}{u}^i \odot \underset{\sim}{v}_i = \omega \leftarrow \text{(scalar)}$$
$$\underset{\sim}{u}^i \otimes \underset{\sim}{v}^i = \underset{\sim}{M}^j \leftarrow \text{(vector)}$$
$$\underset{\sim}{u}^i \odot \underset{\sim}{v}^j = \underset{\sim}{A}^{ij} \leftarrow \text{(Matrix)}$$

The first which is commonly referred to as dot product results in a scalar quantity such as work or energy in the case of multiplication of force and displacement vectors. The second, which is often referred to as the cross product, results in a vector such as moment defined by the multiplication of force and distance vectors. The third one which is introduced by J.W. Gibbs is referred to as dyadic product and results in a matrix.

The magnitude (length or norm) of any vector $\underset{\sim}{u}$ in Euclidean n-space is defined as the square root of the dot product of the vector by itself.

$$|u| = (\underset{\sim}{u}^i \odot \underset{\sim}{u}_i)^{1/2}$$

If $\underset{\sim}{u}^i \odot \underset{\sim}{v}_i = 0$, the two vectors are said to be orthogonal (or perpendicular to each other).

A set of m vectors $(\underset{\sim}{u}_1, \underset{\sim}{u}_2, \ldots, \underset{\sim}{u}_m)$ each with n components is said to be linearly dependent if and only if there exists a set of scalars $(c_1, c_2, \ldots, c_m)$ such that

$$c_1 \underset{\sim}{u}_1 + c_2 \underset{\sim}{u}_2 + \ldots + c_m \underset{\sim}{u}_m = 0$$

Otherwise the vectors are said to be linearly independent. While two dependent vectors define the same direction, three dependent vectors define the same plane. In general, if vectors are linearly dependent, any one of them can be expressed as a linear combination of the others.

The following are the special types of matrices:

Unit matrix (Identity matrix), $\underset{\sim}{I}$ is a square matrix with 1's along the principal diagonal and zero elsewhere.

$$\underset{\sim}{I}\underset{\sim}{A} = \underset{\sim}{A}\underset{\sim}{I} = \underset{\sim}{A}$$
$$\underset{\sim}{I}\underset{\sim}{I} = \underset{\sim}{I}^2 = \underset{\sim}{I}$$
$$\underset{\sim}{I}^n = \underset{\sim}{I}$$

Zero matrix (Null matrix), $\underset{\sim}{0}$ is a matrix which consists of zeros but nothing else.

$$\underset{\sim}{A}\underset{\sim}{0} = \underset{\sim}{0}\underset{\sim}{A} = \underset{\sim}{0}$$

Diagonal matrix, $\underset{\sim}{D}$ is one whose only non-zero elements are those along the principal diagonal.

$$d_{ii} \neq 0, \quad d_{ij} = 0$$

Triangular matrices (Upper and lower), $\underset{\sim}{U}$ and $\underset{\sim}{L}$, are the matrices such that

$$u_{ij} = 0 \quad \text{for} \quad i > j$$
$$\ell_{ij} = 0 \quad \text{for} \quad i < j$$

Symmetric and **skew-symmetric** matrices are the square matrices such that

$$a_{ij} = a_{ji}$$
$$a_{ij} = -a_{ji}$$

respectively.

Band matrix (striped matrix) is one in which all the non-zeo elements are located on or near the main diagonal, that is

$$a_{ij} = 0 \quad \text{for} \quad |i-j| > b$$

Orthogonal matrix is a square matrix whose inverse is equal to its transpose.

$$a_{ik} a_{jk} = \delta_{ij}$$

In other words

$$\underset{\sim}{A}^* = \underset{\sim}{A}^{-1}$$

$$\underset{\sim}{A}\underset{\sim}{A}^* = \underset{\sim}{I}$$

The multiplication of matrices requires special attention. When two matrices $\underset{\sim}{A}$ and $\underset{\sim}{B}$ are multiplied as $\underset{\sim}{A}\underset{\sim}{B}$, matrix $\underset{\sim}{A}$ is called the premultiplier and $\underset{\sim}{B}$ is the postmultiplier. Such a designation should strictly be kept in mind since matrix multiplication is in general non-commutative; in other words

$$\underset{\sim}{A}\underset{\sim}{B} \neq \underset{\sim}{B}\underset{\sim}{A}$$

In order for the product of two matrices to exist, the number of columns of the premultiplier must be equal to the number of rows of the post-multiplier.

$$\underset{(m \times n)}{\underset{\sim}{A}} \quad \underset{(n \times p)}{\underset{\sim}{B}} = \underset{(m \times p)}{\underset{\sim}{C}}$$

where

$$c_{ij} = a_{is} b_{sj} \quad , \quad s = 1, 2, \ldots, n$$

In matrix multiplication, it is likely that $\underset{\sim}{A}\underset{\sim}{B} = \underset{\sim}{0}$ even though $\underset{\sim}{A} \neq 0$ and $\underset{\sim}{B} \neq \underset{\sim}{0}$. Furthermore, if $\underset{\sim}{A}\underset{\sim}{B} = \underset{\sim}{A}\underset{\sim}{C}$, it does not necessarily imply that $\underset{\sim}{B} = \underset{\sim}{C}$.

Interchanging the rows and columns of a matrix $\underset{\sim}{A}$ results in the transpose matrix and is denoted at $\underset{\sim}{A}^*$. For example

$$\underset{\sim}{A} = \begin{bmatrix} 2 & -1 & 0 \\ 1 & 3 & 5 \end{bmatrix} \quad \underset{\sim}{A}^* = \begin{bmatrix} 2 & 1 \\ -1 & 3 \\ 0 & 5 \end{bmatrix}$$

The reader is asked to prove the following properties of transposition:

I. $(\underset{\sim}{A}^*)^* = \underset{\sim}{A}$

II. $(\underset{\sim}{A} \pm \underset{\sim}{B})^* = \underset{\sim}{A}^* \pm \underset{\sim}{B}^*$

III. $(\underline{AB})^* = \underline{B}^* \underline{A}^*$

IV. $\underline{A}^* = \underline{A}$ if and only if $\underline{A}$ is symmetric

V. $\underline{A}^* \underline{A} = \underline{A}\underline{A}^* = \underline{B}$ where $\underline{B}$ is symmetric

VI. $\underline{A} \pm \underline{A}^* = \underline{B}$ where $\underline{B}$ is symmetric

If the determinant of a square matrix vanishes, i.e., if the rows (or columns) are linearly dependent, the matrix is said to be singular. Every singular matrix is also referred to as degenerate matrix. For a square matrix of order n the degree of degeneracy is q when at least one of its q th minors does not vanish. The q th minor of a matrix, on the other hand, is defined as the matrix obtained by omission of any q rows and q columns from the original matrix. Instead of mentioning the degree of degeneracy of a matrix, the reference is often given to its rank. Thus the rank becomes equal to n-q. Therefore, the rank of every square matrix with non-vanishing determinant is the same as its order.

The division law in matrix algebra is not defined; instead, the inversion takes its place.

$$\underline{A}\,\underline{A}^{-1} = \underline{A}^{-1}\,\underline{A} = \underline{I}$$

in which $\underline{A}^{-1}$ is referred to as the inverse or reciprocal of matrix $\underline{A}$. The necessary and sufficient condition for the existance of the inverse is that the original matrix be a non-singular square matrix.

Inversion is one of the most time-consuming operations in matrix algebra. Next to the classical inversion of a matrix by adjoint

$$\underline{A}^{-1} = \frac{adj \cdot \underline{A}}{|\underline{A}|}$$

inversion by operations on rows (columns or both) is very popular.

$$\underline{T}_n \underline{T}_{n-1} \cdots \underline{T}_2 \underline{T}_1 \underline{A} = \underline{I} \quad \text{(row operations)}$$

$$\underline{A}\,\underline{T}_1 \underline{T}_2 \cdots \underline{T}_n \underline{T}_{n-1} = \underline{I} \quad \text{(column operations)}$$

$$\underline{T}_n \underline{T}_{n-1} \cdots \underline{A} \cdots \underline{T}_{r-1} \underline{T}_r = \underline{I} \quad \text{(mixed operations)}$$

Matrices

When the order of the matrix is too large, it may be partitioned into smaller submatrices and the inverse may be obtained in terms of inverses of matrices of lower order.

$$\underset{\sim}{A} = \left[\begin{array}{c|c} \underset{\sim}{A}_{11} & \underset{\sim}{A}_{12} \\ \hline \underset{\sim}{A}_{21} & \underset{\sim}{A}_{22} \end{array} \right]$$

provided that at least one of the submatrices on the main diagonal, $\underset{\sim}{A}_{11}$ or $\underset{\sim}{A}_{22}$, be a non-singular square matrix.

Let the inverse of A be partitioned in the same way.

$$\underset{\sim}{A}^{-1} = \underset{\sim}{B} = \left[\begin{array}{c|c} \underset{\sim}{B}_{11} & \underset{\sim}{B}_{12} \\ \hline \underset{\sim}{B}_{21} & \underset{\sim}{B}_{22} \end{array} \right]$$

From the definition of inverse, i.e., $\underset{\sim}{A}\underset{\sim}{B} = \underset{\sim}{I}$, one may obtain the following:

$$\underset{\sim}{B}_{22} = \underset{\sim}{\Delta}^{-1}$$
$$\underset{\sim}{B}_{21} = -\underset{\sim}{\Delta}^{-1} \underset{\sim}{A}_{21} \underset{\sim}{A}_{11}^{-1}$$
$$\underset{\sim}{B}_{12} = -\underset{\sim}{A}_{11}^{-1} \underset{\sim}{A}_{12} \underset{\sim}{\Delta}^{-1}$$
$$\underset{\sim}{B}_{11} = \underset{\sim}{A}_{11}^{-1} - \underset{\sim}{A}_{11}^{-1} \underset{\sim}{A}_{12} \underset{\sim}{B}_{21}$$

where

$$\underset{\sim}{\Delta} = \underset{\sim}{A}_{22} - \underset{\sim}{A}_{21} \underset{\sim}{A}_{11}^{-1} \underset{\sim}{A}_{12}$$

These steps indicate that the inversion of a matrix of order n by partitioning requires inversion of two matrices of order p and q where $p + q = n$.

Among the many useful properties of the inverse, the following are worth to be proven by the reader.

1. $(\underset{\sim}{A}^{-1})^{-1} = \underset{\sim}{A}$

II. $(ABCD)^{-1} = D^{-1} C^{-1} B^{-1} A^{-1}$

III. $(A^*)^{-1} = (A^{-1})^*$

IV. $(A^{-1})^p = (A^p)^{-1}$

V. $(kA)^{-1} = \dfrac{1}{k} A^{-1}$

VI. $A = \begin{bmatrix} A_{11} & & & \\ & A_{22} & & \\ & & \ddots & \\ & & & A_{nn} \end{bmatrix}, A^{-1} = \begin{bmatrix} \dfrac{1}{A_{11}} & & & \\ & \dfrac{1}{A_{22}} & & \\ & & \ddots & \\ & & & \dfrac{1}{A_{nn}} \end{bmatrix}$

VII. $(A + B)^n = A^n + n A^{n-1} B + \dfrac{n(n-1)}{2!} A^{n-2} B^2 + \cdots$
$+ \dfrac{n(n-1)\cdots(n-p+1)}{p!} A^{n-p} B^p + \cdots + B^n$

When a matrix of larger order is inverted, due to round-off and truncation of numbers, the inverse matrix may not be exact. Let B be the approximate inverse of A such that

$$A B = I - E$$

where E is the error matrix. If A^{-1} represents the true inverse, then

$$B = A^{-1} (I - E)$$
$$A^{-1} = B (I - E)^{-1}$$

According to the property VII (binominal theorem), however,

$$A^{-1} = B + BE + BE^2 + BE^3 + \cdots$$

the convergence is attained provided that the elements of $\underset{\sim}{E}$ are all small, i.e., $\underset{\sim}{B}$ is a fairly accurate inverse of $\underset{\sim}{A}$.

The derivation or integration of matrices is done by differentiating or integrating every element of the matrix in the conventional manner. For example,

$$\frac{\partial}{\partial x} \begin{bmatrix} y^2 x^3 & 4y \\ 2x & -yx^2 \end{bmatrix} = \begin{bmatrix} 3y^2 x^2 & 0 \\ 2 & -2yx \end{bmatrix}$$

Similarly

$$\int \begin{bmatrix} y^2 x^3 & 4y \\ 2x & -yx^2 \end{bmatrix} dx = \begin{bmatrix} \frac{y^2 x^4}{4} & 4yx \\ x^2 & -\frac{1}{3} yx^3 \end{bmatrix} + \begin{bmatrix} c_1 & c_2 \\ c_3 & c_4 \end{bmatrix}$$

It can be shown that

$$\frac{d(\underset{\sim}{A} \pm \underset{\sim}{B})}{dx} = \frac{d\underset{\sim}{A}}{dx} + \frac{d\underset{\sim}{B}}{dx}$$

$$\frac{d(\underset{\sim}{A}\,\underset{\sim}{B})}{dx} = \frac{d\underset{\sim}{A}}{dx} \underset{\sim}{B} + \underset{\sim}{A} \frac{d\underset{\sim}{B}}{dx}$$

Therefore

$$\frac{d\underset{\sim}{A}^2}{dx} \neq 2\underset{\sim}{A} \frac{d\underset{\sim}{A}}{dx}$$

Similarly

$$\frac{d\underset{\sim}{A}^{-1}}{dx} = -\underset{\sim}{A}^{-1} \frac{d\underset{\sim}{A}}{dx} \underset{\sim}{A}^{-1}$$

Furthermore

I. $e^{\underset{\sim}{A}} = \underset{\sim}{I} + \underset{\sim}{A} + \frac{1}{2!} \underset{\sim}{A}^2 + \frac{1}{3!} \underset{\sim}{A}^3 + \cdots$

II. $e^{\underset{\sim}{A}} e^{\underset{\sim}{B}} = e^{\underset{\sim}{A}+\underset{\sim}{B}}$

III. $e^{\underset{\sim}{A}} e^{-\underset{\sim}{A}} = \underset{\sim}{I}^0 = \underset{\sim}{I}$

IV. $\sin \underset{\sim}{A} = \underset{\sim}{A} - \frac{1}{3} \underset{\sim}{A}^2 + \frac{1}{5} \underset{\sim}{A}^5 - \ldots$

V. $\cos \underset{\sim}{A} = \underset{\sim}{I} - \frac{1}{2} \underset{\sim}{A}^2 + \frac{1}{4} \underset{\sim}{A}^4 - \ldots$

1.2. Simultaneous Equations

A set of simultaneous equations can be regarded as a matrix equation in the form of

$$\underset{\sim}{A}\underset{\sim}{X} = \underset{\sim}{B}$$

in which $\underset{\sim}{A}$ is m by n coefficient matrix, X and B are the vectors of unknowns and constants. If $\underset{\sim}{B} = 0$, the set is said to be homogeneous. A nonhomogeneous set can be classified consitent or inconsistent depending upon whether the set possesses a solution or not. A consistent equation on the other hand, may have a unique or an infinite number of solutions. The uniqueness is attained if and only if $\underset{\sim}{A}$ is non-singular.

The solution procedures are basically divided into two categories: direct methods; iterative methods. Gauss' and Cholesky's schemes are the most popular among the direct methods. While Gauss' method aims at the triangularization of the coefficient matrix, Cholesky's decomposes it into two matrices, upper and lower triangular form; then both use back substitution.

Gauss' Elimination:

$$a_{ij} = a_{ij} - \frac{a_{kj}}{a_{kk}} a_{ik} \qquad \begin{array}{l} k = 1, 2, \ldots, n-1 \\ i = k+1, \ldots, n \\ j = k, \ldots, n+1 \end{array}$$

$$x_i = \frac{1}{a_{ii}} (a_{i,n+1} - \sum_{r=i+1}^{n} a_{ir} x_r)$$

where $a_{i,n+1} = b_i$ (the right hand side).

Although this is the simplest and quickest version of Gauss' elimination,

Simultaneous Equations

two serious handicaps may be encountered. The first one is that a_{kk} (the pivot) must be non-zero at each stage since it is a divisor; the second one is that the pivot must be the largest in absolute value otherwise it may cause loss of accuracy in the results. Therefore, in practice, partial or complete pivoting strategy is employed.

Cholesky's Decomposition:

The original equation $\underset{\sim}{A}\underset{\sim}{X} = \underset{\sim}{B}$ can be written as $\underset{\sim}{L}\underset{\sim}{U}\underset{\sim}{X} - \underset{\sim}{B} = 0$ or $\underset{\sim}{L}(\underset{\sim}{U}\underset{\sim}{X} - \underset{\sim}{C}) = 0$ where $\underset{\sim}{L}\underset{\sim}{C} = \underset{\sim}{B}$. Therefore,

$$\ell_{ij} = a_{ij} - \sum_{r=1}^{j-1} \ell_{ir} u_{rj}$$

$$u_{ij} = \frac{1}{\ell_{ii}} (a_{ij} - \sum_{r=1}^{i-1} \ell_{ir} u_{rj})$$

$$x_i = (u_{i,n+1} - \sum_{r=i+1}^{n} u_{ir} x_r)$$

where

$$u_{i,n+1} = c_i$$

In case $\underset{\sim}{A}$ is symmetric i.e., the stiffness and flexibility matrices in descrete elastic systems, then the above $\underset{\sim}{L}\underset{\sim}{U}$ decomposition becomes

$$\underset{\sim}{A} = \underset{\sim}{L}\underset{\sim}{L}^* = \underset{\sim}{U}^*\underset{\sim}{U}$$

thus the method involves square roots. Although for positive definite systems the method does not involve complex numbers, the computer time can be shortened if the square root operations are avoided by the following modifications.

Let

$$u_{ii} = \sqrt{s_{ii}} \quad , \quad u_{ij} = s_{ij} \sqrt{s_{ii}}$$

then

$$s_{ii} = a_{ii} - \sum_{r=1}^{i-1} s_{ri}^2 s_{rr}$$

$$s_{ij} = \frac{1}{s_{ii}} (a_{ij} - \sum_{r=1}^{i-1} s_{ri} s_{rj} s_{rr})$$

$$i = 1, \ldots, n$$
$$j = i+1, \ldots, n+1$$

$$x_i = (s_{i,n+1} - \sum_{r=i+1}^{n} s_{ir} x_r)$$

In systems where the non-zero elements are located around the main diagonal, that is $a_{ij} = 0$ for $|i-j| > \omega-1$, the range of these equations can be modified according to the band width. In doing that, only one half of the band is stored in the machine.

When the system of equation $\underset{\sim}{A}\underset{\sim}{X} = \underset{\sim}{B}$ is in large order and sparse, the iterative methods may be employed for its solution. These methods, in general, are self-correcting and the accuracy of the solution depends upon the number of iterations.

For certain systems, however, they are not convergent. In most cases, rearrangement of equations (conditioning of the set) may produce convergence. For diagonally dominant systems, i.e., $|a_{ii}| > |a_{ij}|$ for instance, the convergent is almost assured. Many sets, in spite of being well-conditioned, depending upon the starting point (initial vector for the unknowns) may very well diverge.

The iterative methods are not suitable for the solutions of equations where the right-hand side contains more than one column. Such a situation arises in the analysis of structures subject to multiple loading conditions.

1.3. Coordinate Transformations

Although most equations in engineering contain physical quantities that are independent of coordinate axes, they are considered incomplete unless the coordinate system in which they are expressed is defined beforehand. For example, a vector representing the displacement of a point becomes meaningless unless the coordinate system associated with it is also defined. Furthermore, most all operations such as additions, multiplications, etc. in matrix algebra are also defined in a common coordinate system. Consequently, the entities involved with such an operation must all be transformed into a common coordinate system. Furthermore, a coordinate transformation may bring some of the entities in an equation to a simpler form. Any operation such as triangularization, diagonalization or elimination may be regarded as a kind of coordinate transformation.

Suppose a set of coordinate axes x_i (linear or non-linear orthogonal or not) is related to another set x_i' by an equation $x_i' = \Psi(x_j)$ where $i, j = 1, 2, \ldots, n$ then, an entity i.e., vector, matrix etc. or an equation expressed in one of these

Coordinate Transformations

coordinate systems can be transformed into another provided that the above relationship is single valued and continuous in the vicinity of the point of the domain where the transformation takes place.

Let the relationship between the two coordinate systems be

$$x'_i = \underset{\sim}{A}_{ij} x_j$$

where

$$\underset{\sim}{A}_{ij} = \frac{\partial x'_i}{\partial x_j} x_j = \underset{\sim}{J} \frac{x'_i}{x_j}$$

which is usually called the Jacobian between the two coordinate systems.

In orthogonal systems, if a right-hand system is transformed into another right-hand system (or left-hand into another left-hand) the determinant of the Jacobian is always + 1; otherwise -1. Matrix $\underset{\sim}{A}$, in general, is called transformation matrix. When a transformation does not include translation; then it is called centro-affine transformation, otherwise general affine transformation. If the elements of matrix $\underset{\sim}{A}$ are not functions of x_i or x'_i such transformation is called linear transformation and assures that the coordinate lines of both systems are straight lines. If at least one of the coordinate systems is curvilinear, the transformation or curvilinear transformation. Such transformations depend upon the point in the domain where the transformation takes place. In certain points in the domain a non-linear coordinate transformation may not even exist. The transformation, however, becomes locally affine if one restricts himself to the immediate vicinity of the point where the transformation takes place.

The coordinate transformations in general are reversible and the inverse transformation is defined as

$$x_j = \underset{\sim}{B}_{ji} x'_i$$

where

$$\underset{\sim}{B}_{ji} = \frac{\partial x_j}{\partial x'_i}$$

and

$$\underset{\sim}{A}_{ij} \underset{\sim}{B}_{jp} = \underset{\sim}{I}$$

Chap. 1. Matrices, Simultaneous Equations and Coordinate Transformations

One of the most common coordinate transformations is the rotational transformation between the orthogonal coordinate axes. In this case, the Jacobian between the two systems represents the directional cosines of the new axes in reference to the old.

$$\underset{\sim}{J} \frac{x'_i}{x_j} = \underset{\sim}{R} = \begin{bmatrix} \ell_1 & m_1 & n_1 \\ \ell_2 & m_2 & n_2 \\ \ell_3 & m_3 & n_3 \end{bmatrix}$$

Pre-multiplication of any vectors defined in the old coordinate system by $\underset{\sim}{R}$ results in the same vector expressed in the new system.

$$\underset{\sim}{V'} = \underset{\sim}{R} \underset{\sim}{V}$$

The inverse transformation, can be done by multiplying both sides with $\underset{\sim}{R}$

$$\underset{\sim}{V} = \underset{\sim}{R}^{-1} \underset{\sim}{V'}$$

In orthogonal systems, however,

$$\underset{\sim}{V} = \underset{\sim}{R}^{*} \underset{\sim}{V'}$$

since $\underset{\sim}{R}^{-1} = \underset{\sim}{R}^{*}$

Finally a matrix of order (m × n) can be regarded as n vectors in m-space or m vectors in n-space, and the elements in each row (column) may be interpreted as the components of the corresponding row (column) vectors. Consequently, the transformation of a vector from one coordinate frame into another becomes

$$\underset{\sim}{A'} = \underset{\sim}{R} \underset{\sim}{A} \underset{\sim}{R}^{-1}$$

or

$$\underset{\sim}{A'} = \underset{\sim}{R} \underset{\sim}{A} \underset{\sim}{R}^{*}$$

for orthogonal systems.

Chapter 2
INTRODUCTION TO STIFFNESS AND FLEXIBILITY METHODS

Basically there are two different types of matrix methods to analyze structures, namely, stiffness (displacement) and flexibility (force) methods which are also known as equilibrium and compatibility methods. The third method (the mixed method) which is the combination of the two will not be presented here.

Both methods satisfy the force equilibrium equations and the displacement compatibility conditions but not in the same order. In the stiffness method the force equilibriums and in the flexibility method the displacement compatibilities are satisfied first. The choice of one method over the other depends upon the structure as well as the preference of the analyst. Each method eventually involves the solution of simultaneous equations in which the nodal displacements are the unknown quantities in the stiffness methods, member forces (stresses) in the flexibility method and partly nodal displacements partly member forces in the mixed method.

2.1. Stiffness Method

Let an elastic body be subject to a set of forces P_{it} applied in a quasi-linear manner. If Δ_{it} represent the displacement of point i in the direction of P_{it} at the time of t, then the work done by these forces after the loading process is completed is

$$U = \frac{1}{2}(P_1 \Delta_1 + P_2 \Delta_2 + \cdots + P_n \Delta_n) \qquad (2.1)$$

where P_i and Δ_i are the final values of loads and displacements respectively.

Now, suppose that a small variation to one of the displacements, say Δ_i, is introduced; then by the chain rule of differentiation, the variation of the total strain energy in respect to Δ_i becomes

$$\frac{\partial U}{\partial \Delta_i} = \frac{1}{2}\left[\frac{\partial P_1}{\partial \Delta_i}\Delta_1 + \frac{\partial P_2}{\partial \Delta_i}\Delta_2 + \cdots \frac{\partial P_i}{\partial \Delta_i}\Delta_i + P_i \cdots + \frac{\partial P_n}{\partial \Delta_i}\Delta_n\right] \qquad (2.2)$$

According to Castigliano's first theorem, however,

$$\frac{\partial U}{\partial \Delta_i} = P_i \qquad (2.3)$$

from which

(2.4) $$\underset{\sim}{P}_i = \frac{\partial \underset{\sim}{P}_1}{\partial \underset{\sim}{\Delta}_i} \underset{\sim}{\Delta}_1 + \frac{\partial \underset{\sim}{P}_2}{\partial \underset{\sim}{\Delta}_i} \underset{\sim}{\Delta}_2 + \cdots + \frac{\partial \underset{\sim}{P}_n}{\partial \underset{\sim}{\Delta}_i} \underset{\sim}{\Delta}_n$$

If i varies from 1 to n this equation can be written in a matrix form as

(2.5) $$\begin{bmatrix} \underset{\sim}{P}_1 \\ \underset{\sim}{P}_2 \\ \vdots \\ \underset{\sim}{P}_n \end{bmatrix} = \begin{bmatrix} \dfrac{\partial \underset{\sim}{P}_1}{\partial \underset{\sim}{\Delta}_1} & \dfrac{\partial \underset{\sim}{P}_2}{\partial \underset{\sim}{\Delta}_1} & \cdots & \dfrac{\partial \underset{\sim}{P}_n}{\partial \underset{\sim}{\Delta}_1} \\ \dfrac{\partial \underset{\sim}{P}_1}{\partial \underset{\sim}{\Delta}_2} & \dfrac{\partial \underset{\sim}{P}_2}{\partial \underset{\sim}{\Delta}_2} & \cdots & \dfrac{\partial \underset{\sim}{P}_n}{\partial \underset{\sim}{\Delta}_2} \\ \vdots & & & \vdots \\ \dfrac{\partial \underset{\sim}{P}_1}{\partial \underset{\sim}{\Delta}_n} & \dfrac{\partial \underset{\sim}{P}_2}{\partial \underset{\sim}{\Delta}_n} & \cdots & \dfrac{\partial \underset{\sim}{P}_n}{\partial \underset{\sim}{\Delta}_n} \end{bmatrix} \begin{bmatrix} \underset{\sim}{\Delta}_1 \\ \underset{\sim}{\Delta}_2 \\ \vdots \\ \underset{\sim}{\Delta}_n \end{bmatrix}$$

The left-hand side of this equation represents external forces applied on the system (including reactions), and the column matrix on the right represents displacements of the nodal points (including those at the supports). If an element of the square matrix in this equation is written as

$$\frac{\partial \underset{\sim}{P}_i}{\partial \underset{\sim}{\Delta}_j} = \frac{\partial (\partial U / \partial \underset{\sim}{\Delta}_i)}{\partial \underset{\sim}{\Delta}_j} = \frac{\partial^2 U}{\partial \underset{\sim}{\Delta}_j \partial \underset{\sim}{\Delta}_i} = \frac{\partial \underset{\sim}{P}_j}{\partial \underset{\sim}{\Delta}_i}$$

the symmetry will be observed. Furthermore such an element represents the holding force required (or developed) at point i in order to keep the body in equilibrium when a unit displacement is introduced at point i.

Equation (2.5) is also valid for an elastic system if there are only two nodal points (a line element).

Stiffness Method

$$\begin{bmatrix} \underset{\sim}{P_1} \\ \cdots \\ \underset{\sim}{P_2} \end{bmatrix} = \begin{bmatrix} \dfrac{\partial P_1}{\partial \Delta_1} & \vdots & \dfrac{\partial P_2}{\partial \Delta_1} \\ \cdots & \vdots & \cdots \\ \dfrac{\partial P_1}{\partial \Delta_2} & \vdots & \dfrac{\partial P_2}{\partial \Delta_2} \end{bmatrix} \begin{bmatrix} \Delta_1 \\ \cdots \\ \Delta_2 \end{bmatrix}$$

If such an element is labeled as ij element, the above equation becomes

$$\begin{bmatrix} \underset{\sim}{P_{ij}} \\ \cdots \\ \underset{\sim}{P_{ji}} \end{bmatrix} = \begin{bmatrix} \underset{\sim}{K_{ii}^j} & \vdots & \underset{\sim}{K_{ij}} \\ \cdots & \vdots & \cdots \\ \underset{\sim}{K_{ji}} & \vdots & \underset{\sim}{K_{jj}^i} \end{bmatrix} \begin{bmatrix} \underset{\sim}{\Delta_{ij}} \\ \cdots \\ \underset{\sim}{\Delta_{ji}} \end{bmatrix} \qquad (2.6)$$

which is known as the stiffness matrix equation of member ij. It represents forces developed at the ends of a memeber from the end displacements.

Now, considering that member end displacements and the nodal point displacements are the same (compatibility condition)

$$\underset{\sim}{\Delta_i} = \underset{\sim}{\Delta_{ij}} = \underset{\sim}{\Delta_{ia}} = \cdots\cdots = \underset{\sim}{\Delta_{in}} \qquad (2.7)$$

and that the sum of member-end forces of all members joining at a joint is equal to the external load applied to that joint (equilibirum condition)

$$\underset{\sim}{P_i} = \underset{\sim}{P_{ij}} + \underset{\sim}{P_{im}} + \cdots\cdots \underset{\sim}{P_{in}} \qquad (2.8)$$

the equilibirum of nodal point i becomes

$$\underset{\sim}{P_i} = \underset{\sim}{K_{ii}}\,\underset{\sim}{\Delta_i} + \underset{\sim}{K_{ij}}\,\underset{\sim}{\Delta_j} + \cdots\cdots + \underset{\sim}{K_{in}}\,\underset{\sim}{\Delta_n} \qquad (2.9)$$

where

$$\underset{\sim}{K_{ii}} = \underset{\sim}{K_{ii}^j} + \underset{\sim}{K_{ii}^m} + \cdots\cdots + \underset{\sim}{K_{ii}^n}$$

Notice the similarities between Eqs. (2.5) and (2.9). Assuming that i can take any value from 1 to n, Eq. (2.9) becomes

$$(2.10) \quad \begin{bmatrix} P_1 \\ P_2 \\ \vdots \\ P_n \end{bmatrix} = \begin{bmatrix} K_{11} & K_{12} & \cdots & K_{1n} \\ & K_{22} & & K_{2n} \\ & & \ddots & \vdots \\ & & & K_{nn} \end{bmatrix} \begin{bmatrix} \Delta_1 \\ \Delta_2 \\ \vdots \\ \Delta_n \end{bmatrix}$$

or in short

$$P = K \Delta$$

This is the complete stiffness matrix equation (another version of Eq. 5) of the entire body prior to the boundary conditions. Once the boundary conditions are introduced, the solution of this equation yields the unknown displacements as the free nodal points. The elements on the main diagonal of this matrix represent the direct stiffness of nodal points and the diagonal elements represent the cross-stiffness between the nodal points.

2.2. Flexibility Method

This method, like the stiffness method, establishes the force-displacement relationship. While in the stiffness method forces are expressed in terms of displacements, in the flexibility method displacements are expressed in terms of forces. Therefore, the formulation of this method follows different paths. According to Castigliano's second theorem, for instance, Eq. (2.1) takes the following form:

$$\frac{\partial U}{\partial P_i} = \Delta_i = \frac{1}{2} \left[P_1 \frac{\partial \Delta_1}{\partial P_i} + P_2 \frac{\partial \Delta_2}{\partial P_i} + \cdots + \Delta_i + P_i \frac{\partial \Delta_i}{\partial P_i} + \cdots P_n \frac{\partial \Delta_n}{\partial P_i} \right]$$

from which

$$\Delta_i = P_1 \frac{\partial \Delta_1}{\partial P_i} + P_2 \frac{\partial \Delta_2}{\partial P_i} + \cdots P_n \frac{\partial \Delta_n}{\partial P_i}$$

Flexibility Method

Considering the variation of i from 1 to n, then,

$$\begin{bmatrix} \Delta_1 \\ \Delta_2 \\ \vdots \\ \Delta_n \end{bmatrix} = \begin{bmatrix} \dfrac{\partial \Delta_1}{\partial P_1} & \dfrac{\partial \Delta_2}{\partial P_1} & \cdots & \dfrac{\partial \Delta_n}{\partial P_1} \\[4pt] \dfrac{\partial \Delta_1}{\partial P_2} & \dfrac{\partial \Delta_2}{\partial P_2} & \cdots & \dfrac{\partial \Delta_n}{\partial P_2} \\[4pt] \vdots & \vdots & & \vdots \\[4pt] \dfrac{\partial \Delta_1}{\partial P_n} & \dfrac{\partial \Delta_2}{\partial P_n} & \cdots & \dfrac{\partial \Delta_n}{\partial P_n} \end{bmatrix} \begin{bmatrix} P_1 \\ P_2 \\ \vdots \\ P_n \end{bmatrix} \qquad (2.11)$$

or, shortly,

$$\Delta = D\, P \qquad (2.12)$$

This equation represents the complete flexibility matrix equation of the system prior to the application of the boundary conditions.

The elements of D in this equation become meaningless unless the body is stabilized for the variation of forces. If a body is subject to a system of forces in equilibrium and any one of these forces is altered while all the others are kept constant (implication of partial derivation), the body will undergo infinite displacement.

Stabilization is achieved by setting a sufficient number of displacements (dependencies of vectors) equal to zero.

$$\frac{\partial \Delta_j}{\partial P_i} = 0$$

which yields the primary structure. The boundary conditions which are similar to this restriction are introduced later.

The assembly of D starts with the equilibrium of joints

$$P_i = C\, P_{ij} \qquad (2.13)$$

where P_i is the applied loads (known) at the free joints and P_{ij} forces developed at the ends of member ij (unknown). The rectangular matrix C becomes a non-singular square matrix for statically determinant systems--in which case the physical properties of members need not be considered. The solution of Eq. (2.13) is the result of the analysis. In case $\underset{\sim}{C}$ is not square, it can be partitioned as

(2.14) $$\underset{\sim}{P}_I = \underset{\sim}{C}_I \underset{\sim}{P}_{ij} + \underset{\sim}{C}_{II} \underset{\sim}{P}_{II}$$

in which $\underset{\sim}{P}_I$, $\underset{\sim}{P}_{ij}$ and $\underset{\sim}{P}_{II}$ respectively, represent the known (given) forces on the free joints, the unknown member forces and the redundants. Furthermore, let the strain energy of the system be written as

(2.15) $$U = \frac{1}{2} \Sigma \underset{\sim}{P}_{ij}^* \underset{\sim}{D}_{ij} \underset{\sim}{P}_{ij}$$

where $\underset{\sim}{D}_{ij}$ is the flexibility matrix of member ij such that $\underset{\sim}{D}_{ij}$ times $\underset{\sim}{P}_{ij}$ results in $\underset{\sim}{e}_{ij}$ (the deformation of member ij). Equation (2.14) can be written such that the left-hand side would contain all member forces including redundants.

(2.16) $$\begin{bmatrix} \underset{\sim}{P}_{ij} \\ \underset{\sim}{P}_{II} \end{bmatrix} = \underset{\sim}{B}_I \underset{\sim}{P}_I + \underset{\sim}{B}_{II} \underset{\sim}{P}_{II}$$

where

$$\underset{\sim}{B}_I = \begin{bmatrix} \underset{\sim}{C}_I^{-1} \\ 0 \end{bmatrix} \qquad \underset{\sim}{B}_{II} = \begin{bmatrix} -\underset{\sim}{C}_I^{-1} \underset{\sim}{C}_{II} \\ I \end{bmatrix}$$

If Eq. (2.16) is substituted into Eq. (2.15), the strain energy of the system becomes

(2.17) $$U = \frac{1}{2} \begin{bmatrix} \underset{\sim}{P}_I^* & \underset{\sim}{P}_{II}^* \end{bmatrix} \underset{\sim}{D} \begin{bmatrix} \underset{\sim}{P}_I \\ \underset{\sim}{P}_{II} \end{bmatrix}$$

in which

(2.18) $$\underset{\sim}{D} = \begin{bmatrix} \underset{\sim}{B}_I^* \\ \underset{\sim}{B}_{II}^* \end{bmatrix} \begin{bmatrix} \underset{\sim}{D}_{12} & & \\ & \underset{\sim}{D}_{ij} & \\ & & \underset{\sim}{D}_{mn} \end{bmatrix} \begin{bmatrix} \underset{\sim}{B}_I & \underset{\sim}{B}_{II} \end{bmatrix}$$

represents the same complete flexibility matrix presented in Eq. (2.11). The diagonal matrix in Eq. (2.18) is referred to as the unassembled flexibility matrix of the structure and contains the flexibility matrices of each member in the global coordinate system.

Substituting Eq. (2.18) into Eq. (2.11) and introducing the boundary conditions, result in the final equation; then the solution of it yields the forces developed in each element. The introduction of the boundary conditions is presented in the next article.

Chapter 3
BOUNDARY CONDITIONS

There are three conditions that every structural problem must comply with. They are: compatibility, equilibrium and boundary conditions. The first two are taken care of by Eqs. (2.7) and (2.8) respectively.

The boundary conditions are specified in terms of forces and displacements. The given forces form the generalized force vector in the complete matrix equation of the systems (Eqs. 2.5 and 2.11). Notice that these equations are of mixed nature containing known and unknown quantities on both sides of the equality (the unknown reactions in P and the unknown displacements in Δ). Therefore, the set must be rearranged according to the given pattern of the supports.

3.1. In Stiffness Method

The supports can be either unyielding ($\Delta_i = 0$), elastic ($\Delta_i = \Psi P_i$), or constant ($\Delta_i = C$). Let a problem possess all these supports. Further assume that the governing equation is partitioned as

$$(3.1) \quad \begin{bmatrix} P_1 \\ P_2 \\ P_3 \\ P_4 \end{bmatrix} = \begin{bmatrix} K_{11} & K_{12} & K_{13} & K_{14} \\ K_{21} & K_{22} & K_{23} & K_{24} \\ K_{31} & K_{32} & K_{33} & K_{34} \\ K_{41} & K_{42} & K_{43} & K_{44} \end{bmatrix} \begin{bmatrix} \Delta_1 = 0 \\ \Delta_2 = \Psi P_2 \\ \Delta_3 = C \\ \Delta_4 = ? \end{bmatrix}$$

from which one may obtain the following for the unknown nodal point displacements

$$(3.2) \quad \Delta_4 = \left\{ I - K_{44}^{-1} K_{42} \Psi G^{-1} K_{24} \right\}^{-1} \left\{ K_{44}^{-1} [P_4 - K_{43} \Psi G^{-1} K_{23} C - K_{43} C] \right\}$$

in which

$$G = [I - K_{22} \Psi]$$

Ψ = spring constants (linear or non-linear)

$\underset{\sim}{C}$ = support movements

Once the free joint displacements are known, the reactions at the supports can be calculated from the following equations.

$$\underset{\sim}{P}_1 = \underset{\sim}{K}_{22} \underset{\sim}{\Psi} \underset{\sim}{G}^{-1} [\underset{\sim}{K}_{24} \underset{\sim}{\Delta}_4 + \underset{\sim}{K}_{23} \underset{\sim}{C}] + \underset{\sim}{K}_{23} \underset{\sim}{C} + \underset{\sim}{K}_{24} \underset{\sim}{\Delta}_4$$

$$\underset{\sim}{P}_2 = \underset{\sim}{G}^{-1} [\underset{\sim}{K}_{24} \underset{\sim}{\Delta}_4 + \underset{\sim}{K}_{23} \underset{\sim}{C}]$$

$$\underset{\sim}{P}_3 = \underset{\sim}{K}_{32} \underset{\sim}{\Psi} \underset{\sim}{G}^{-1} [\underset{\sim}{K}_{24} \underset{\sim}{\Delta}_4 + \underset{\sim}{K}_{23} \underset{\sim}{C}] + \underset{\sim}{K}_{33} \underset{\sim}{C} + \underset{\sim}{K}_{34} \underset{\sim}{\Delta}_4$$

in which $\underset{\sim}{P}_1$, $\underset{\sim}{P}_2$, $\underset{\sim}{P}_3$ respectively represent reactions in unyielding, spring and settled supports.

Notice that the solution of the problem requires inversion of matrices of order $\underset{\sim}{K}_{22}$ and $\underset{\sim}{K}_{44}$. In other words, in the stiffness method, among the two structures with the same number of nodal points, the one that is highly supported requires inversion of matrices of lower order. For instance, assume that the structure is supported by unyielding supports only, i.e., $\underset{\sim}{\Psi} = \underset{\sim}{C} = \underset{\sim}{0}$. Then, the solution is achieved by inversion of $\underset{\sim}{K}_{44}$ whose order is equal to the number of free nodal points.

$$\underset{\sim}{\Delta}_4 = \underset{\sim}{K}_{44}^{-1} \underset{\sim}{P}_4$$
$$\underset{\sim}{P}_1 = \underset{\sim}{K}_{24} \underset{\sim}{\Delta}_4 \tag{3.3}$$

3.2. In Flexibility Method

The above partioning technique for the introduction of boundary conditions can equally be applied in the flexibility method. Let, for instance, a structure have only unyielding supports. The complete flexibility equation (Eq. 2.11) for this structure can be partitioned as

$$\begin{bmatrix} \underset{\sim}{\Delta}_1 = \underset{\sim}{0} \\ \hline \underset{\sim}{\Delta}_4 = ? \end{bmatrix} = \begin{bmatrix} \underset{\sim}{D}_{11} & | & \underset{\sim}{D}_{14} \\ \hline \underset{\sim}{D}_{41} & | & \underset{\sim}{D}_{44} \end{bmatrix} \begin{bmatrix} \underset{\sim}{P}_1 = ? \\ \hline \underset{\sim}{P}_4 \end{bmatrix}$$

which yields

(3.4) $$\underset{\sim}{P_1} = \underset{\sim}{D_{11}^{-1}} \underset{\sim}{D_{14}} \underset{\sim}{P_4}$$

for the reactions (redundants) and

(3.4a) $$\underset{\sim}{\Delta_4} = \underset{\sim}{D_{41}} \underset{\sim}{D_{11}^{-1}} \underset{\sim}{D_{14}} + \underset{\sim}{D_{44}} \underset{\sim}{P_4}$$

for the unknown displacements at the free joints.

The comparison of Eqs. (3.3) and (3.4) indicates that for highly indeterminate systems the stiffness method is advantageous over the flexibility method because $\underset{\sim}{K_{44}} < \underset{\sim}{D_{11}}$ in size. When the number of supports of a structure increases, the order of $\underset{\sim}{K_{44}}$ decreases and that of $\underset{\sim}{D_{11}}$ increases. The size of these two matrices is often considered in chosing the method of analysis for the structure.

The redundants in statically indeterminate structures can be internal member forces as well as an excess number of reactions. In either case, however, displacements in the direction of redundants (relative in the first case and absolute in the latter) are always zero. Equation (3.4), therefore, can be obtained directly from the strain energy of the system by using Castigliano's second theorem. The strain energy can be expressed as

$$U = \frac{1}{2} \begin{bmatrix} \underset{\sim}{P_1^*} & \vdots & \underset{\sim}{P_4^*} \end{bmatrix} \begin{bmatrix} \underset{\sim}{\Delta_1} \\ \hdashline \underset{\sim}{\Delta_4} \end{bmatrix}$$

or

$$U = \frac{1}{2} \begin{bmatrix} \underset{\sim}{P_1^*} & \vdots & \underset{\sim}{P_4^*} \end{bmatrix} \underset{\sim}{D} \begin{bmatrix} \underset{\sim}{P_1} \\ \hdashline \underset{\sim}{P_4} \end{bmatrix}$$

from which

$$\frac{\partial U}{\partial \underset{\sim}{P_1}} = \underset{\sim}{\Delta_1} = \frac{1}{2} \left\{ \begin{bmatrix} \underset{\sim}{I} & \vdots & \underset{\sim}{0} \end{bmatrix} \underset{\sim}{D} \begin{bmatrix} \underset{\sim}{P_1} \\ \hdashline \underset{\sim}{P_4} \end{bmatrix} + \begin{bmatrix} \underset{\sim}{P_1} & \underset{\sim}{P_4} \end{bmatrix} \underset{\sim}{D} \begin{bmatrix} \underset{\sim}{I} \\ \hdashline \underset{\sim}{0} \end{bmatrix} \right\}$$

or

$$\underset{\sim}{\Delta_1} = \begin{bmatrix} \underset{\sim}{I} & \vdots & \underset{\sim}{0} \end{bmatrix} \underset{\sim}{D} \begin{bmatrix} \underset{\sim}{P_1} \\ \hdashline \underset{\sim}{P_4} \end{bmatrix}$$

which represents the displacements of the points of application of the redundants in the direction of the redundants. That, of course, is always zero. Therefore

$$[\underset{\sim}{I} \mid \underset{\sim}{0}] \left[\begin{array}{c|c} \underset{\sim}{D_{11}} & \underset{\sim}{D_{14}} \\ \hline \underset{\sim}{D_{41}} & \underset{\sim}{D_{44}} \end{array}\right] \left[\begin{array}{c} \underset{\sim}{P_1} \\ \hline \underset{\sim}{P_4} \end{array}\right] = \underset{\sim}{0}$$

or

$$\underset{\sim}{P_1} = - \underset{\sim}{D_{11}^{-1}} \underset{\sim}{D_{14}} \underset{\sim}{P_4}$$

which is identical with Eq. (3.4).

Note that

$$\frac{\partial \underset{\sim}{U}}{\partial \underset{\sim}{P_1}} = \underset{\sim}{0}$$

is known as the principle of minimum strain energy which advocates that only true values of redundants make the strain energy minimum.

Sometimes the boundary conditions are not always prescribed in the direction of the global coordinate axes. Since the complete equation (Eq. 3.1) is written in reference to a common (global) coordinate system, the introduction of zero displacements at the boundaries can not be done by simply deleting the corresponding rows and columns. Let, for instance, certain displacement components of nodal point 1 be prescribed zero in the direction of a primed coordinate system which is different than the global system. Further assume that $\underset{\sim}{R_1}$ is the rotation matrix between these two coordinate systems such that

$$\underset{\sim}{\Delta_1'} = \underset{\sim}{R_1} \underset{\sim}{\Delta_1}$$
$$\underset{\sim}{P_1'} = \underset{\sim}{R_1} \underset{\sim}{P_1}$$

If, now, the complete equation is written in the following partitioned form

$$\left[\begin{array}{c} \underset{\sim}{P_1} \\ \hline \underset{\sim}{P_2} \end{array}\right] = \left[\begin{array}{c|c} \underset{\sim}{K_{11}} & \underset{\sim}{K_{12}} \\ \hline \underset{\sim}{K_{21}} & \underset{\sim}{K_{22}} \end{array}\right] \left[\begin{array}{c} \underset{\sim}{\Delta_1} \\ \hline \underset{\sim}{\Delta_2} \end{array}\right]$$

then the transformation of $\underset{\sim}{\Delta}$, from the global to the primed coordinate system can be done by

$$\left[\begin{array}{c} \underset{\sim}{P'_1} \\ \hline \underset{\sim}{P_2} \end{array}\right] = \left[\begin{array}{c|c} \underset{\sim}{R_1}\underset{\sim}{K_{11}}\underset{\sim}{R_1^*} & \underset{\sim}{R_1}\underset{\sim}{K_{12}} \\ \hline \underset{\sim}{K_{21}}\underset{\sim}{R_1^*} & \underset{\sim}{K_{22}} \end{array}\right] \left[\begin{array}{c} \underset{\sim}{\Delta'_1} \\ \hline \underset{\sim}{\Delta_2} \end{array}\right]$$

in which certain components of $\underset{\sim}{\Delta}$, are equal to zero. Consequently, the corresponding rows and columns can now be deleted from the above equation. If there are more than one inclined supports, they can all be transformed into the primed coordinate axes by multiplying the corresponding rows and columns with their proper rotation matrices.

In summary, the modification of the complete stiffness (flexibility) matrix equation of a structure that contains an inclined support (s) at its i th joint is done by premultiplying its i th row by $\underset{\sim}{R_i}$ and post-multiplying its i th column by $\underset{\sim}{R_i^*}$. The boundary conditions, then, are introduced as usual by erasing the corresponding rows and columns from the complete equation.

Chapter 4
ANALYSIS OF MODIFIED SYSTEMS

The analysis of elastic systems is done for the purpose of determining whether or not stresses and deflections developed throughout the body meet certain requirements. Such requirements are referred to as the design criteria. If safe and economical designs can be achieved by means other than the analysis, there would not be any need for their analysis. At the end of the first analysis, it is quite likely to be found out that certain portions of the system are either too weak or too strong. This in turn necessitates the modification, i.e., changing properties of the elements, and the reanalysis of the system. Sometimes, modification of member sizes alone may not be sufficient to achieve the optimum design. Changing the boundary conditions, adding or deleting members, modifying the geometry of the system may all be needed. All of these require the reanalysis of the structure.

Now there are two alternatives: either to start the analysis from scratch — in a way, analyzing a new system — or to take advantage of the work done prior to the modification. If the former is preferred, (certainly it is not advised) there would be no need for the material presented in this chapter. Otherwise, it will be shown that the amount of work in reanalyzing the next modified structure will diminish in each cycle as the modifications decrease.

The other theories which are capable of doing cycles automatically will not be mentioned here as they must all comply with certain limitations.

Besides the changes in geometry, the modifications can be divided into two categories: those on the boundaries and those on the physical properties of the elements. We shall discuss the two separately.

4.1. Modification at the boundaries

Assume that the system is already analyzed under one set of boundary conditions and later on the same system is to be analyzed under another set of boundary conditions. Further assume that the modified system has a greater number of restrictions at the boundaries. Let the final stiffness matrix equation $P = K \Delta$ and its solution $\Delta = D P$ of the original structure be partioned as

$$\begin{bmatrix} P_I \\ \hline P_{II} \end{bmatrix} = \begin{bmatrix} K_{I,I} & \vdots & K_{I,II} \\ \cdots & \cdots & \cdots \\ K_{II,I} & \vdots & K_{II,II} \end{bmatrix} \begin{bmatrix} \Delta_I \\ \hline \Delta_{II} \end{bmatrix} \quad (4.1)$$

and

(4.2)
$$\begin{bmatrix} \underset{\sim}{\Delta}_I \\ \cdots \\ \underset{\sim}{\Delta}_{II} \end{bmatrix} = \begin{bmatrix} \underset{\sim}{D}_{I,I} & \vdots & \underset{\sim}{D}_{I,II} \\ \cdots & \cdots & \cdots \\ \underset{\sim}{D}_{II,I} & \vdots & \underset{\sim}{D}_{II,II} \end{bmatrix} \begin{bmatrix} \underset{\sim}{P}_I \\ \cdots \\ \underset{\sim}{P}_{II} \end{bmatrix}$$

in which $\underset{\sim}{P}_{II}$ and $\underset{\sim}{\Delta}_{II}$ represent forces and displacements respectively at those boundaries which are subject to modifications.

If now the absolute restrictions are exposed on $\underset{\sim}{\Delta}_{II}$, i.e., $\underset{\sim}{\Delta}_{II} = \underset{\sim}{0}$, then the analysis of the modified system would look like

$$\begin{bmatrix} \underset{\sim}{\Delta}'_I \\ \cdots \\ \underset{\sim}{0} \end{bmatrix} = \begin{bmatrix} \underset{\sim}{D}_{I,I} & | & \underset{\sim}{D}_{I,II} \\ \cdots & & \cdots \\ \underset{\sim}{D}_{II,I} & | & \underset{\sim}{D}_{II,II} \end{bmatrix} \begin{bmatrix} \underset{\sim}{P}_I \\ \cdots \\ \underset{\sim}{P}'_{II} \end{bmatrix}$$

where $\underset{\sim}{\Delta}'_I$ and $\underset{\sim}{P}'_{II}$ are the modified free joint displacements and the forces developed at the new supports. Both of these vectors can be obtained from the results of the original analysis are

(4.3)
$$\underset{\sim}{\Delta}'_I = \underset{\sim}{D}_{I,I} \underset{\sim}{P}_I + \underset{\sim}{D}_{I,II} \underset{\sim}{P}'_{II}$$
$$\underset{\sim}{P}'_{II} = \underset{\sim}{D}^{-1}_{II,II} \underset{\sim}{D}_{II,I} \underset{\sim}{P}_I$$

Finally

$$\underset{\sim}{\Delta}'_I = \underset{\sim}{D}_{I,I} \underset{\sim}{P}_I - \underset{\sim}{D}_{I,II} \underset{\sim}{D}^{-1}_{II,II} \underset{\sim}{D}_{II,I} \underset{\sim}{P}_I$$

or

(4.4)
$$\underset{\sim}{\Delta}'_I = [\underset{\sim}{D}_{I,I} - \underset{\sim}{D}_{I,II} \underset{\sim}{D}^{-1}_{II,II} \underset{\sim}{D}_{II,I}] \underset{\sim}{P}_I$$

This equation indicates that the analysis of a modified structure makes use of the results (inverse matrix) of the original structure and requires inversion of a relatively small matrix $\underset{\sim}{D}_{II,II}$.

Now consider the kind of boundary modification in which the modified system becomes less rigid (fewer restrictions at the boundaries) than the original structure. Since the final equation of the original system is smaller in size than that of the modified system, it will not contain all the information needed for the modified structure. For instance,

Modification at the Boundaries

$$\Delta_I = D_{I,I}\, P_I \begin{bmatrix} \Delta'_I \\ \Delta'_{II} \end{bmatrix} = \begin{bmatrix} D_{I,I} & \vdots & D_{I,II} \\ D_{II,I} & \vdots & D_{II,II} \end{bmatrix} \begin{bmatrix} P_I \\ P_{II} \end{bmatrix} \quad (4.5)$$

represent the governing equations of the original and the modified systems respectively.

Considering that

$$K\, D = I$$

for the modified structure, and

$$K_{I,I}^{-1} = D_{I,I}$$

as results of the modified structure, Eq. (4.5) yields the following results:

$$\begin{aligned}
\Delta'_I &= D_{I,I}\,[\, P_I - K_{I,II}\, D_{II,II}\, P_{II}\,] \\
\Delta'_{II} &= D_{II,II}\,[\, K_{II,I}\, D_{I,I}\, P_I + P_{II}\,]
\end{aligned} \quad (4.6)$$

where

$$\begin{aligned}
D_{II,II} &= [\, K_{II,II} - K_{II,I}\, K_{I,I}^{-1}\, K_{I,II}\,]^{-1} \\
D_{I,II} &= -\, D_{II,I}^{*} = -\, K_{I,I}^{-1}\, K_{I,II}\, D_{II,II}
\end{aligned}$$

These results again indicate that the displacements of the modified structure can be obtained by inverting a relatively small matrix.

4.2. Modification on Member Properties

In order to meet the design requirements, the changes in the physical properties of members are very common at the end of each analysis. Such modification normally decreases as the design advances.

Let

$$\begin{bmatrix} P_I \\ \hline P_{II} \end{bmatrix} = \begin{bmatrix} K_{I,I} + K'_{I,I} & \vdots & K_{I,II} \\ \hline K_{II,I} & \vdots & K_{II,II} \end{bmatrix} \begin{bmatrix} \Delta'_I \\ \hline \Delta'_{II} \end{bmatrix} \quad (4.7)$$

represent the final stiffness matrix equation of the modififed structure which differs from the original equation in the amount of $\underset{\sim}{K}'_{I,I}$ representing the algebraic differences between the old and the new stiffness matrices of the modified members.

If Eq. (4.7) is re-written as

(4.8)
$$\left[\begin{array}{c} \underset{\sim}{P}_I \\ \hline \underset{\sim}{P}_{II} \end{array}\right] - \left[\begin{array}{c|c} \underset{\sim}{K}'_{I,I} \underset{\sim}{\Delta}'_I \\ \hline 0 \end{array}\right] = \left[\begin{array}{c|c} \underset{\sim}{K}_{I,I} & \underset{\sim}{K}_{I,II} \\ \hline \underset{\sim}{K}_{II,I} & \underset{\sim}{K}_{II,II} \end{array}\right] \left[\begin{array}{c} \underset{\sim}{\Delta}'_I \\ \hline \underset{\sim}{\Delta}'_{II} \end{array}\right]$$

and is multiplied by the inverse of the original stiffness matrix one may obtain

(4.9)
$$\left[\begin{array}{c} \underset{\sim}{\Delta}'_I \\ \hline \underset{\sim}{\Delta}'_{II} \end{array}\right] = \left[\begin{array}{c|c} \underset{\sim}{D}_{I,I} & \underset{\sim}{D}_{I,II} \\ \hline \underset{\sim}{D}_{II,I} & \underset{\sim}{D}_{II,II} \end{array}\right] \left[\begin{array}{c} \underset{\sim}{P}_I - \underset{\sim}{K}'_{I,I} \underset{\sim}{\Delta}'_I \\ \hline \underset{\sim}{P}_{II} \end{array}\right]$$

from which

$$\underset{\sim}{\Delta}'_I = [\underset{\sim}{I} + \underset{\sim}{D}_{I,I} \underset{\sim}{K}'_{I,I}]^{-1} [\underset{\sim}{D}_{I,I} \underset{\sim}{P}_I + \underset{\sim}{D}_{I,II} \underset{\sim}{P}_{II}]$$
$$\underset{\sim}{\Delta}'_{II} = \underset{\sim}{D}_{II,I} \left\{\underset{\sim}{P}_I - \underset{\sim}{K}'_{I,I} [\underset{\sim}{I} + \underset{\sim}{D}_{I,I} \underset{\sim}{K}'_{I,I}]^{-1} [\underset{\sim}{D}_{I,I} \underset{\sim}{P}_I + \underset{\sim}{D}_{I,II} \underset{\sim}{P}_{II}]\right\} + \underset{\sim}{D}_{II,II} \underset{\sim}{P}_{II}$$

Calling

$$[\underset{\sim}{I} + \underset{\sim}{D}_{I,I} \underset{\sim}{K}'_{I,I}]^{-1} = \overline{\underset{\sim}{D}}_{I,I}$$

and remembering that

$$\underset{\sim}{D}_{I,I} \underset{\sim}{P}_I + \underset{\sim}{D}_{I,II} \underset{\sim}{P}_{II} = \underset{\sim}{\Delta}_I$$
$$\underset{\sim}{D}_{II,I} \underset{\sim}{P}_I + \underset{\sim}{D}_{II,II} \underset{\sim}{P}_{II} = \underset{\sim}{\Delta}_{II}$$

the results become

(4.10)
$$\underset{\sim}{\Delta}'_I = \overline{\underset{\sim}{D}}_{I,I} \underset{\sim}{\Delta}_I$$
$$\underset{\sim}{\Delta}'_{II} = -\underset{\sim}{D}_{II,I} \underset{\sim}{K}'_{I,I} \underset{\sim}{\Delta}'_I + \underset{\sim}{\Delta}_{II}$$

The time and labor needed to obtain the displacements of the modified system in this way depend upon the number of elements modified which in turn determine the size of $\underset{\sim}{K}'_{I,I}$. Since such a number is often relatively small, the method presented here may be found very advantageous.

Chapter 5
ANALYSIS OF LARGE SYSTEMS BY METHOD OF TEARING

When a system is too large to be handled as a whole (the final stiffness or flexibility matrix exceeds the main memory capacity of the machine), there are two things to be done; both involve partitioning. The first one is to partition the stiffness/flexibility matrix and operate on the sub-matrices one at a time; the second one is to partition the physical system into smaller units (sub-systems) and analyze each unit independent by satisfying the force and displacement compatibilities at the intersections where the partitioning takes place.

The latter was first introduced by Gabriel Kron to the analysis of electric circuits and elastic structure in the form of sub-spaces. Later on, Kron named his method Diakoptics meaning tearing apart in Greek, also referred to as the method of sub-structures in structural engineering.

Considering that the elastic systems can be symmetric or non-symmetric, they should be treated accordingly.

5.1. Non-Symmetric Systems

In general, a structure could be partitioned into any number of units and the partitioning may take place at any location. However, since the labor will increase as the number of sub-structures increases, it is advisable to keep these numbers as small as possible. In other words, no partitioning is recommended unless it is necessary. It would also be advisable to arrange relatively the same size sub-systems and to make the separation by cutting fewer elements.

Consider the structure shown in Fig. 5.1. Assume that this structure is separated into two sub-structures. Each structure will be in equilibrium under its own external forces and the unknown internal forces at the interfaces.

Let the final stiffness matrix equation (after the introduction of natural boundary conditions) of each system be arranged as

$$\begin{bmatrix} \underset{\sim}{P}_I^A \\ \cdots \\ \underset{\sim}{P}_{II}^A \end{bmatrix} = \begin{bmatrix} \underset{\sim}{K}_{I,I}^A & | & \underset{\sim}{K}_{I,II}^A \\ ---&+&--- \\ \underset{\sim}{K}_{II,I}^A & | & \underset{\sim}{K}_{II,II}^A \end{bmatrix} \begin{bmatrix} \underset{\sim}{\Delta}_I^A \\ --- \\ \underset{\sim}{\Delta}_{II}^A \end{bmatrix}$$

Non-Symmetric Systems

and

$$\begin{bmatrix} P^B_{\sim I} \\ \cdots \\ P^B_{\sim II} \end{bmatrix} \begin{bmatrix} K^B_{\sim I,I} & \vdots & K^B_{\sim I,II} \\ \cdots & \cdots & \cdots \\ K^B_{\sim II,I} & \vdots & K^B_{\sim II,II} \end{bmatrix} \begin{bmatrix} \Delta^B_{\sim I} \\ \cdots \\ \Delta^B_{\sim II} \end{bmatrix} \quad (5.1)$$

in which

$P^A_{\sim I}$, $P^B_{\sim I}$ are the known external forces on systems A and B

$P^A_{\sim II}$, $P^B_{\sim II}$ are the unknown forces at the interfaces

$\Delta^A_{\sim I}$, $\Delta^B_{\sim I}$ are the unknown displacements in systems A and B

$\Delta^A_{\sim II}$, $\Delta^B_{\sim II}$ are the unknown displacements at the interfaces

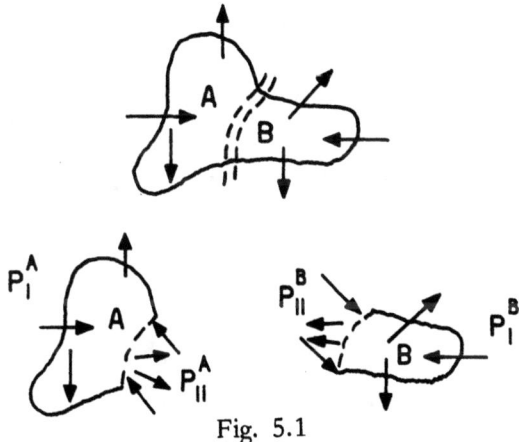

Fig. 5.1

Notice that neither one of the above two sets has a unique solution, i.e., neither contains more unknowns than the number of equations. However, the following relationships (which are referred to as force and displacement compatibilities) between the two sets

$$P^A_{\sim II} = - P^B_{\sim II} \quad , \quad \Delta^A_{\sim II} = \Delta^B_{\sim II}$$

make the total number of equations equal to the total number of unknowns.

Now it is quite possible that the square matrices in Eq. (5.1) may not have any inversion. Such a thing occurs when one of the sub-systems is kinematically unstable. Since the partitioning is often done without paying any attention to kinematic, i.e., cutting a tall structure by mid-height which results in the upper portion being kinematically unstable, the inversion procedure is not advised toward the solution of these sets. Consequently, assume that through an elimination procedure the first set in Eq. (5.1) is brought to the following form:

$$(5.2) \quad \begin{bmatrix} \underset{\sim}{P}_I \\ \cdots \\ \underset{\sim}{P}^A_{II} + \underset{\sim}{\bar{P}}^A_{II} \end{bmatrix} = \begin{bmatrix} \bar{K}^A_{I,I} & \vdots & \bar{K}^A_{I,II} \\ \cdots & \vdots & \cdots \\ \underset{\sim}{0} & \vdots & \bar{K}^A_{II,II} \end{bmatrix} \begin{bmatrix} \Delta^A_I \\ \cdots \\ \Delta^A_{II} \end{bmatrix}$$

where $\bar{P}^A_{II}$ represents the modification of P^A_{II} during the elimination and is a function of P^A_I. Note that such an elimination is possible in spite of the fact that the set has no solution.

If the last portion of Eq. (5.2) is written as

$$P^A_{II} = \bar{K}^A_{II,II} \Delta^A_{II} - \bar{P}^A_{II} = -P^B_{II}$$

and substituted into the second set of Eq. (5.1), one may obtain the following:

$$(5.3) \quad \begin{bmatrix} P^B_I \\ \cdots \\ \bar{P}^A_{II} \end{bmatrix} = \begin{bmatrix} K^B_{I,I} & \vdots & K^B_{I,II} \\ \cdots & \vdots & \cdots \\ K^B_{II,I} & \vdots & K^B_{II,II} + \bar{K}^A_{II,II} \end{bmatrix} \begin{bmatrix} \Delta^B_I \\ \cdots \\ \Delta^B_{II} \end{bmatrix}$$

The square matrix on the right of this equation is the same as the final stiffness matrix of structure B except for its right principal sub-matrix $K^B_{II,II}$ which is modified in the amount of $\bar{K}^A_{II,II}$. Also the load vector on the left is modified in the amount of $\bar{P}^A_{II}$ which is a function of the forces exerted on structure A. The solution of this equation yields the free joint displacements in structure B including those at the interfaces. Complying with the compatibilities, if the interface

Symmetric Systems

displacements are substituted into the first set of Eq. (5.1), the results would be the displacements of structure A. The stresses and strain throughout the body, finally, can be determined once the displacements are known.

The method can be extended to such cases where there are more than two sub-systems. It is also worth remembering that whether any one of the sub-systems is kinematically stable or not makes no difference.

5.2. Symmetric Systems

We shall now consider systems that a least possess one symmetry axis. The majority of structures, i.e., buildings, bridges, ships, airplanes, cars etc., are symmetrical. There are certain facts and principles that hold for symmetrical systems which are not valid otherwise. It is, therefore, proper to give special consideration to the properties of symmetrical structures and to make use of them in their analysis. Such structures can be separated though their symmetry axis into sub-systems and the analysis may be performed in parts. In doing this, certain boundary conditions will be imposed at the interfaces such that no further consideration need be given to the effect of one sub-structure on the others. Therefore, separation of symmetrical structures into sub-systems differs from that presented in the previous article for general structures. In other words, in the previous article, the location of separation was arbitrary, whereas here it is restricted to the symmetry axes only.

Consider the symmetrical structure shown in Fig. 5.2. which is subject to a general loading. According to the principle of superposition, the load on this structure can be divided into two loadings, a symmetrical one as shown in (b) and an antisymmetrical one as shown in (c).

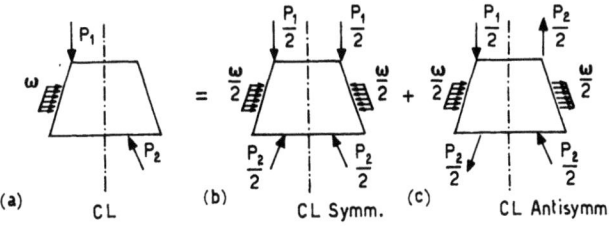

Fig. 5.2

Note that such a division could be done for any type of loading.

Consider now the structure in (b) where the structure and the loading are both symmetrical. It would then be reasonable to say that the deformed shape of this structure would also be symmetrical. This in turn conditions that the deflections of points on the symmetry axis could be in the direction of this axis. In structure (b), on the other hand, the points on the symmetry axis will not be displaced in the direction of that axis. Consequently, only one half of the original structure need be analyzed under the boundary conditions shown in Fig. 5.3.

Fig. 5.3

The analysis of two sub-structures in this fashion may not seem to have any advantages; nevertheless it is advantageous and such an advantage becomes more appreciable as the structure gets larger.

Especially one pays attention to the fact that after the analysis of either one of the two sub-structures, the next one can be analyzed in shorter time by the modification of the previous analysis according to the method presented in Chapter 4.

Chapter 6
EQUIVALENT STIFFNESS–FLEXIBILITY MATRICES FOR SERIES, PARALLEL AND CLOSED–LOOP SYSTEMS

Quite often, in practice, systems contain series and parallel connected members or closed loops. Such special arrangements of members occur, for instance, in building structures, pipe lines, continuous bridges, etc. The analysis of these systems, then, can be expedited and the results may be improved if they are replaced (theoretically) by a single member with an equivalent flexibility (stiffness). In doing this, with the exception of the end points, all the intermediate nodal points will be eliminated. This in turn reduces the size of the overall stiffness (flexibility) matrix of the entire system.

6.1. Members in Series

Although we shall here treat only line elements, the method is also applicable to other types of elements. Consider the system shown in Fig. 6.1. where a few members are connected in series.

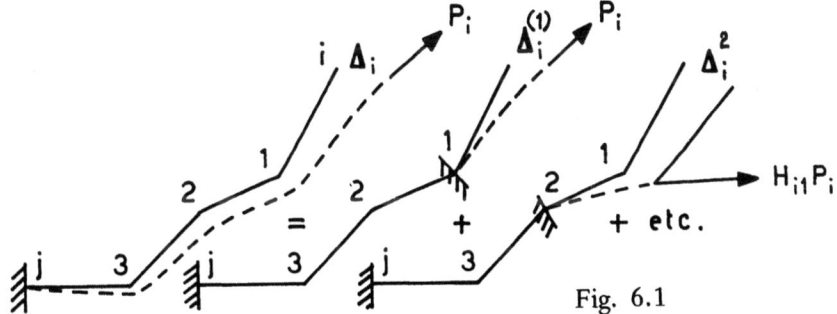

Fig. 6.1

By the definition of "flexibility" (variation of displacement in respect to force), the equivalent flexibility matrix of this system could be obtained by introducing forces at i and evaluating displacements developed there. Granting that this can be done by direct integration from i to j, nevertheless, we shall obtain it by the summation of element flexibilities from i to j.

First of all, the deflection of point i caused by forces introduced at i would be

$$\underset{\sim}{\Delta}_i^{(1)} = \underset{\sim}{D}_{i\ell} \underset{\sim}{P}_i$$

The next thing would be the translation of forces from i to point 1, then the translation of deflections from point 1 back to i.

$$\underset{\sim}{\Delta}_i^{(2)} = \underset{\sim}{H}_{i\ell}^* D_{12} \underset{\sim}{H}_{i\ell} \underset{\sim}{P}_i$$

Continuing the same way, the final deflections at i would be

(6.1)
$$\underset{\sim}{\Delta}_i = \Sigma \underset{\sim}{H}_{in}^* \underset{\sim}{D}_{n,n+1} \underset{\sim}{H}_{in} \underset{\sim}{P}_i$$

in which

(6.2)
$$\Sigma \underset{\sim}{H}_{in}^* \underset{\sim}{D}_{n,n+1} \underset{\sim}{H}_{in} = \Sigma \underset{\sim}{B}_n = \underset{\sim}{D}_{ij}^e$$

is the equivalent flexibility matrix of the system between i and j, and $\underset{\sim}{B}_n$ is the modified flexibility matrix of each element. The equivalent stiffness is the inverse of $\underset{\sim}{D}_{ii}^e$.

6.2. Members in Parallel

When members are in parallel, the equivalent stiffness matrix can be obtained more easily than its equivalent flexibility.

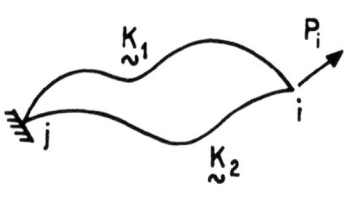

Fig. 6.2

Let $\underset{\sim}{P}_1$ and $\underset{\sim}{P}_2$ represent forces developed at i-end of members 1 and 2 due to forces applied at i. Since the force-displacement relationship of each line is

$$\underset{\sim}{P}_1 = \underset{\sim}{K}_1 \underset{\sim}{\Delta}_i$$
$$\underset{\sim}{P}_2 = \underset{\sim}{K}_2 \underset{\sim}{\Delta}_i$$

and the equilibrium of joint i is

$$\underset{\sim}{P}_i = \underset{\sim}{P}_1 + \underset{\sim}{P}_2$$

therefore

(6.3)
$$\underset{\sim}{P}_i = [\underset{\sim}{K}_1 + \underset{\sim}{K}_2] \underset{\sim}{\Delta}_i$$

from which the equivalent stiffness matrix of the system would be

(6.4)
$$\underset{\sim}{K}_{ij}^e = \Sigma \underset{\sim}{K}_n$$

considering that there might be n number of members in series. The equivalent flexibility matrix, then, would be the inverse of $\underset{\sim}{K}_{ij}^e$.

6.3. Closed Loops

From the results of the previous two articles, it is quite evident that the equivalent flexibility (stiffness) matrices of closed loops requires no special effort. For instance, consider the system shown in Fig. 6.3.

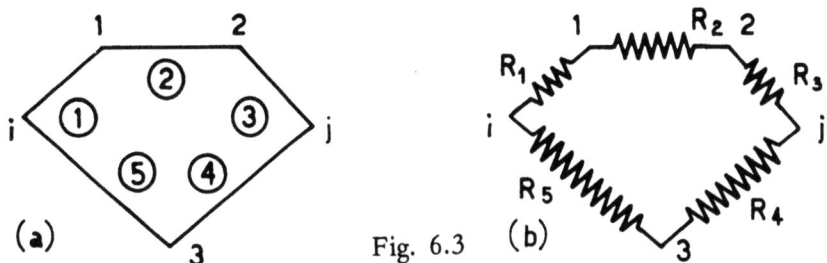

Fig. 6.3

The equivalent stiffness matrix between i and j of this closed loop can be obtained as

$$\underset{\sim}{K}^e_{ij} = [\underset{\sim}{B_1} + \underset{\sim}{B_2} + \underset{\sim}{B_3}]^{-1} + [\underset{\sim}{B_4} + \underset{\sim}{B_5}]^{-1} \qquad (6.5)$$

Similarly, the equivalent stiffness matrix between nodal points 1 and 3, for instance, would be

$$\underset{\sim}{K}^e_{13} = [\underset{\sim}{B_2} + \underset{\sim}{B_3} + \underset{\sim}{B_4}]^{-1} + [\underset{\sim}{B_5} + \underset{\sim}{B_1}]^{-1} \qquad (6.6)$$

The equivalent flexibilities, of course, would be the inverse of the above entities.

Now assume that the entities in Eqs. 6.5 and 6.6 are scalar quantities instead of matrices. Designating them with new parameters as

$$V_{ij} = [R_1 + R_2 + R_3]^{-1} + [R_4 + R_5]^{-1} \stackrel{?}{=} \frac{\Sigma R}{[R_1 + R_2 + R_3][R_4 + R_5]} \qquad (6.5a)$$

$$V_{13} = \frac{\Sigma R}{[R_2 + R_3 + R_4][R_5 + R_1]}$$

would represent voltage drops between points i, j and 1, 3 respectively of the electric circuit shown in Fig. 6.3b. In other words, the well-known analogy between elastic systems and electric circuits is encountered.

$\underset{\sim}{B}$ (Flexibility) = R (Resistance)
$\underset{\sim}{P}$ (Force) = I (Current)
$\underset{\sim}{\Delta}$ (Displacement) = V (Volt).

REFERENCES

[1] Schwartz, J.T., Introduction to Matrices and Vectors. McGraw-Hill, New York, 1961.

[2] Ayres, F. Jr., Theory and Problems of Matrices. Schaum Publishing Co., New York, 1962.

[3] Pipes, L.A., Matrix Methods for Engineering. Prentice-Hall, Englewood Cliffs, N.J., 1963.

[4] Asplund, S.O., "Inversion of Band Matrices", 2nd ASCE Conf. Electronic Computations, 1960.

[5] Gatewood, B.E. and N. Ohanian, "Tro-Diagonal Matrix Method for Complex Structures", J. Structural Division, ASCE 91, No. ST2, 1965.

[6] Norris, C.H. and J.B. Wilbur, Elementary Structural Analysis. McGraw-Hill, New York, 1960.

[7] Timoshenko, S.P. and D.H. Young, Theory of Structures, 2nd ed. McGraw-Hill, New York, 1965.

[8] Argyris, J.H. and S. Kelsey, Energy Theorems and Structural Analysis. Butterworth Scientiific Publications, London, 1960.

[9] Clough, R.W., "The Finite Element Method in Plane Stress Analysis", Proc. Am. Soc. of Civil Engrs. 87, 1960.

[10] Hall, A.S. and R.W. Woodhead, Frame Analysis. Wiley, New York, 1961.

[11] Pei, L.M., "Stiffness Method of Rigid Frame Analysis", Proc. 2nd ASCE Conf. Electronic Computations, 1960.

[12] Pestel, E.C. and F.A. Leckie, Matrix Methods in Elasto-Mechanics. McGraw-Hill, New York, 1963.

[13] Prziemiecki, J.S., "Matrix Structural Analysis of Substructures", AIAA J.1, No. 1, 1963.

[14] Livesley, R.K., Matrix Methods of Structural Analysis. Pergamon Press, Oxford, 1964.

[15] Gallagher, R.H., A Correlation Study of Methods of Matrix Structural Analysis. Pergamon Press, Oxford, 1964.

[16] Gere, J.M. and W. Weaver Jr., Analysis of Framed Structures. Van Nostrand, Princeton, N.J., 1965.

[17] Jones, R.E., "A Generalization of the Direct-Stiffness Mehtod of Structural Analysis", AIAA J. 2, No. 5, 1964.

[18] Fraeijs de Veubeke, B., "Upper and Lower Bounds in Matrix Structural Analysis", AGARDOGRAPH 72, Pergamon Press, London, 1964.

[19] Clough, R.W., E.L. Wilson, and I.P. King, "Large Capacity Multi-story Frame Analysis Programs", J. Structural Division, ASCE 89, No. ST4, 1963.

[20] Proc. Conf. on Matrix Methods in Structural Mechanics, AFFDL-TR-66-80, Wright Patterson Air Force Base, 1965.

[21] Rubinstein, M.F., Matrix Computer Analysis of Structures. Prentice-Hall, Englewood Cliffs, N.J., 1966.

[22] Pian, T.H.H., "Formulations of Finite Element Methods for Solid Continuous", in Recent Advances in Matrix Methods of Structural Analysis and Design. R.H. Gallagher, Y. Yamanda, J.T. Oden (eds.), University of Alabama Press, 1971.

References

[23] Sander, G. and P. Beckers, 4th Conference on Matrix Methods in Structural Mechanics, Wright Patterson Air Force Base, Ohio, "Improvements of Finite Element Solutions for Structural and non-Structural Applications", 1971.

[24] Zienkiewicz, O.C., The Finite Element in Engineering Science. McGraw-Hill, London, 1971.

CONTENTS

	Page
Preface	3
Chapter 1: Matrices, Simultaneous Equations and Coordinate Transformations	5
1.1. Matrices	5
1.2. Simultaneous Equations	12
1.3. Coordinate Transformations	14
Chapter 2: Introduction to Stiffness and Flexibility Methods	17
2.1. Stiffness Method	17
2.2. Flexibility Method	20
Chapter 3: Boundary Conditions	24
3.1. In Flexibility Method	24
3.2. In Stiffness Method	25
Chapter 4: Analysis of Modified Systems	29
4.1. Modification at the Boundaries	29
4.2. Modification on Member Properties	31
Chapter 5: Analysis of Large Systems by Method of Tearing	34
5.1. Non-Symmetric Systems	34
5.2. Symmetric Systems	37
Chapter 6: Equivalent Stiffness-Flexibility Matrices for Series, Parallel and Closed-Loop Systems	39
6.1. Members in Series	39
6.2. Members in Parallel	40
6.3. Closed Loops	41
References	43
Contents	47

MIX
Papier aus verantwortungsvollen Quellen
Paper from responsible sources
FSC® C105338

If you have any concerns about our products,
you can contact us on
ProductSafety@springernature.com

In case Publisher is established outside the EU,
the EU authorized representative is:
**Springer Nature Customer Service Center GmbH
Europaplatz 3, 69115 Heidelberg, Germany**

Printed by Libri Plureos GmbH
in Hamburg, Germany